THE COLLAPSE OF WESTERN CIVILIZATION

Choice manifests itself in society in small increments and moment-to-moment decisions as well as in loud dramatic struggles.

—Lewis Mumford,
Technics and Civilization (1934)

THE COLLAPSE OF WESTERN CIVILIZATION

A View from the Future

Naomi Oreskes and Erik M. Conway

COLUMBIA UNIVERSITY PRESS

NEW YORK

Columbia University Press
Publishers Since 1893
New York   Chichester, West Sussex
cup.columbia.edu

Library of Congress Cataloging-in-Publication Data
Oreskes, Naomi.
The collapse of western civilization : a view from the future /
Naomi Oreskes and Erik M. Conway.
pages cm
Includes bibliographical references.
ISBN 978-0-231-16954-7 (pbk. : alk. paper) —
ISBN 978-0-231-53795-7 (ebook)
1. Civilization, Western—Forecasting. 2. Civilization,
Western—21st century. 3. Science and civilization.
4. Progress—Forecasting. 5. Twenty-first century—Forecasts.
I. Conway, Erik M., 1965– II. Title.
CB158.O64 2014
909'.09821—dc23

2013048899

∞

Columbia University Press books are printed on permanent and
durable acid-free paper.
This book is printed on paper with recycled content.
Printed in the United States of America

COVER DESIGN: Milenda Nan Ok Lee
COVER ART: Colin Anderson © Getty Images

This book is based on the essay of the same name that was originally
published in *Daedalus* (Winter 2013), the journal of The American
Academy of Arts and Sciences. That essay has been slightly
expanded and modified from its original publication, and
the lexicon and interview are new to this book.

Contents

CONTENTS

Acknowledgments

We are grateful to Robert Fri, Stephen Ansolabehere, and the staff at the American Academy of Arts and Sciences for commissioning the original version of this work; to the Institute of Advanced Studies at the University of Western Australia where that version was first written; and to Patrick Fitzgerald, Roy Thomas, Milenda Lee, and the diligent and creative team at Columbia University Press for turning it into a book.

We also thank our agent, Ayesha Pande, without whom our work would be written but not necessarily read; Kim Stanley Robinson for inspiration; and the audience member at the Sydney Writers' Festival who asked one of us: "Will you write fiction next?"

Introduction

Science fiction writers construct an imaginary future; historians attempt to reconstruct the past. Ultimately, both are seeking to understand the present. In this essay, we blend the two genres to imagine a future historian looking back on a past that is our present and (possible) future. The occasion is the tercentenary of the end of Western culture (1540–2093); the dilemma being addressed is how we— the children of the Enlightenment—failed to act on robust information about climate change and knowledge of the damaging events that were about to unfold. Our historian concludes that a second Dark Age had fallen on Western civilization, in which denial and self-deception, rooted in an ideological fixation on "free" markets, disabled the world's powerful nations in the face of tragedy. Moreover, the scientists who best understood the problem were hamstrung by their own cultural practices, which demanded an excessively stringent standard for accepting claims of any kind—even those involving imminent threats. Here, our future historian, living in the Second People's Republic of China, recounts the events of the Period of the Penumbra (1988–2093) that led to the Great Collapse and Mass Migration (2073–2093).

THE COLLAPSE OF WESTERN CIVILIZATION

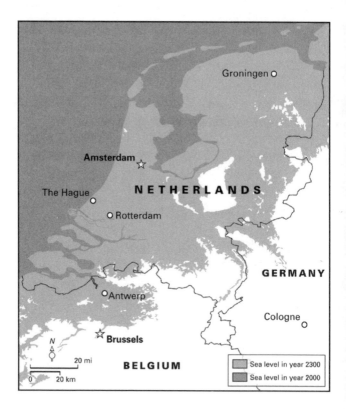

Groningen ○

Amsterdam ☆

The Hague ○

NETHERLANDS

○ Rotterdam

GERMANY

○ Antwerp

Cologne ○

☆ Brussels

N

20 mi

0   20 km

BELGIUM

Sea level in year 2300
Sea level in year 2000

The nation formerly known as the Netherlands Once referred to as the "Low Countries" of Europe, much of the land area of this nation had been reclaimed from the sea by extensive human effort from the sixteenth through the twentieth centuries. The unexpectedly rapid rise of the seas of the Great Collapse overwhelmed the Dutch citizens. The descendants of their survivors largely reside in the Nordo-Scandinavian Union, while the rusting skyscrapers of their drowned cities are a ghostly reminder of a glorious past.

1

The Coming of the Penumbral Age

In the prehistory of "civilization," many societies rose and fell, but few left as clear and extensive an account of what happened to them and why as the twenty-first-century nation-states that referred to themselves as *Western civilization*. Even today, two millennia after the collapse of the Roman and Mayan empires and one millennium after the end of the Byzantine and Inca empires, historians, archaeologists, and synthetic-failure paleoanalysts have been unable to agree on the primary causes of those societies' loss of population, power, stability, and identity. The case of Western civilization is different because the consequences of its actions were not only predictable, but predicted. Moreover, this technologically transitional society left extensive records both in twentieth-century-style paper and in twenty-first-century electronic formats, permitting us to reconstruct what happened in extraordinarily clear detail. While analysts differ on the exact circumstances, virtually all agree that the people of Western civilization knew what was happening to them but were unable to stop it. Indeed, the most startling aspect of this story is just how much these people knew, and how unable they were to act upon what they knew. Knowledge did not translate into power.

For more than one hundred years before its fall, the Western world knew that carbon dioxide (CO_2) and water vapor absorbed heat in the planetary atmosphere. A three-phase Industrial Revolution led to massive release of additional CO_2, initially in the United Kingdom (1750–1850); then in Germany, the United States, the rest of Europe, and Japan (1850–1980); and finally in China, India, and Brazil (1980–2050). (Throughout this essay, I will use the nation-state terms of the era; for the reader not familiar with the political geography of Earth prior to the Great Collapse, the remains of the United Kingdom can be found in present-day Cambria; Germany in the Nordo-Scandinavian Union; and the United States and Canada in the United States of North America.) At the start of the final phase, in the mid-twentieth century, some physical scientists—named as such due to the archaic Western convention of studying the physical world in isolation from social systems—recognized that the anthropogenic increment of CO_2 could theoretically warm the planet. Few were concerned; total emissions were still quite low, and in any case, most scientists viewed the atmosphere as an essentially unlimited sink. Through the 1960s, it was often said that "the solution to pollution is dilution."

Things began to change as planetary sinks approached saturation and "dilution" was shown to be insufficient. Some chemical agents had extremely powerful effects even at very low concentrations, such as organochlorine insecticides (most famously the pesticide dichlorodiphenyltrichloroethane, or DDT) and chlorinated fluorocarbons (CFCs). The former were shown in the 1960s to disrupt reproductive function in fish, birds, and mammals; scientists correctly predicted in the 1970s that the latter would deplete the stratospheric ozone layer. Other saturation effects occurred because of the huge volume of materials being released into the planetary environment. These materials included sulfates from coal combustion, as well as CO_2 and methane (CH_4) from a host of sources including fossil fuel combustion, concrete manufacture, deforestation, and then-prevalent agricultural techniques, such as growing rice in paddy fields and producing cattle as a primary protein source.

In the 1970s, scientists began to recognize that human activities were changing the physical and biological functions of the planet in consequential ways—giving rise to the Anthropocene Period of geological history.[1] None of the scientists who made these early discoveries was particularly visionary: many of the relevant studies were by-products of nuclear weapons testing and development.[2] It was the rare man—in those days, sex discrimination was still widespread—who understood that he was in fact studying the limits of planetary sinks. A notable exception was the futurist Paul Ehrlich, whose book *The Population Bomb* was widely read in the late 1960s but was considered to have been discredited by the 1990s.[3]

Nonetheless, enough research accumulated to provoke some response. Major research programs were launched and new institutions created to acknowledge and investigate the issue. Culturally, celebrating the planet was encouraged on an annual Earth Day (as if every day were not an Earth day!), and in the United States, the establishment of the Environmental Protection Agency formalized the concept of *environmental protection*. By the late 1980s, scientists had recognized that concentrations of CO_2 and other greenhouse gases were having discernible effects on planetary climate, ocean chemistry, and biological systems, threatening grave consequences if not rapidly controlled. Various groups and individuals began to argue for the need to limit greenhouse gas emissions and begin a transition to a non-carbon-based energy system.

In the 1970s, scientists began to recognize that human activities were changing the physical and biological functions of the planet in consequential ways—giving rise to the Anthropocene Period of geological history.

Historians view 1988 as the start of the Penumbral Period. In that year, world scientific and political leaders created a new, hybrid scientific-governmental organization, the Intergovernmental Panel on Climate Change (IPCC), to communicate relevant science and form the foundation for international governance to protect the planet and its denizens. A year later, the Montreal Protocol to Control Substances that Deplete the Ozone Layer became a model for international governance to protect the atmosphere, and in 1992, based on that model, world nations signed the United Nations Framework Convention on Climate Change (UNFCCC) to prevent "dangerous anthropogenic interference" in the climate system. The world seemed to recognize the crisis at hand, and was taking steps to negotiate and implement a solution.

But before the movement to change could really take hold, there was backlash. Critics claimed that the scientific uncertainties were too great to justify the expense and inconvenience of eliminating greenhouse gas emissions, and that any attempt to solve the problem would cost more than it was worth. At first, just a handful of people made this argument, almost all of them from the United States. (In hindsight, the self-justificatory aspects of the U.S. position are obvious, but they were not apparent to many at the time.) Some countries tried but failed to force the United States into international cooperation. Other nations used inertia in the United States to excuse their own patterns of destructive development.

By the end of the millennium, climate change denial had spread widely. In the United States, political leaders—including the president, members of Congress, and members of state legislatures—took denialist positions. In Europe, Australia, and Canada, the message of "uncertainty" was promoted by industrialists, bankers, and some political leaders. Meanwhile, a different version of denial emerged in non-industrialized nations, which argued that the threat of climate change was being used to prevent their development. (These claims had much less environmental impact, though, because these countries produced few greenhouse gas emissions and generally had little international clout.)

There were notable exceptions. China, for instance, took steps to control its population and convert its economy to non-carbon-based energy sources. These efforts were little noticed and less emulated in the West, in part because Westerners viewed Chinese population control efforts as immoral, and in part because the country's exceptionally fast economic expansion led to a dramatic increase in greenhouse gas emissions, masking the impact of renewable energy. By 2050, this impact became clear as China's emissions began to fall rapidly. Had other nations followed China's lead, the history recounted here might have been very different.[4]

But as it was, by the early 2000s, dangerous anthropogenic interference in the climate system was under way. Fires, floods, hurricanes, and heat waves began to intensify. Still, these effects were discounted. Those in what we might call *active denial* insisted that the extreme weather events reflected natural variability, despite a lack of evidence to support that claim. Those in *passive denial* continued life as they had been living it, unconvinced that a

compelling justification existed for broad changes in industry and infrastructure. The physical scientists studying
these steadily increasing disasters did not help quell this
denial, and instead became entangled in arcane arguments
about the "attribution" of singular events. Of course the
threat to civilization inhered not in any individual flood,
heat wave, or hurricane, but in the overall shifting climate
pattern, its impact on the cryosphere, and the increasing
acidification of the world ocean. But scientists, trained as
specialists focused on specific aspects of the atmosphere,
hydrosphere, cryosphere, or biosphere, found it difficult to
articulate and convey this broad pattern.

The year 2009 is viewed as the "last best chance" the
Western world had to save itself, as leaders met in Copenhagen, Denmark, to try, for the fifteenth time since the
UNFCCC was written, to agree on a binding, international law to prevent disruptive climate change. Two years
before, scientists involved
in the IPCC had declared
anthropogenic warming to
be "unequivocal," and public
opinion polls showed that
a majority of people—even
in the recalcitrant United
States—believed that action
was warranted. But shortly

> By the early 2000s, dangerous
> anthropogenic interference in
> the climate system was under
> way. Fires, floods, hurricanes,
> and heat waves began to inten-
> sify. Still, these effects were
> discounted.

before the meeting, a massive campaign was launched to
discredit the scientists whose research underpinned the

IPCC's conclusion. This campaign was funded primarily by fossil fuel corporations, whose annual profits at that time exceeded the GDPs of most countries.[5] (At the time, most countries still used the archaic concept of a *gross domestic product*, a measure of consumption, rather than the Bhutanian concept of gross domestic happiness to evaluate well-being in a state.) Public support for action evaporated; even the president of the United States felt unable to move his nation forward.

Meanwhile, climate change was intensifying. In 2010, record-breaking summer heat and fires killed more than 50,000 people in Russia and resulted in more than $15 billion (in 2009 USD) in damages. The following year, massive floods in Australia affected more than 250,000 people. In 2012, which became known in the United States as the "year without a winter," winter temperature records, including for the highest overnight lows, were shattered— something that should have been an obvious cause for concern. A summer of unprecedented heat waves and loss of livestock and agriculture followed. The "year without a winter" moniker was misleading, as the warm winter was largely restricted to the United States, but in 2023, the infamous "year of perpetual summer" lived up to its name, taking 500,000 lives worldwide and costing nearly $500 billion in losses due to fires, crop failure, and the deaths of livestock and companion animals.

The loss of pet cats and dogs garnered particular attention among wealthy Westerners, but what was anomalous in 2023 soon became the new normal. Even then, political, business, and religious leaders refused to accept that what lay behind the increasing destructiveness of these disasters was the burning of fossil fuels. More heat in the atmosphere meant more energy had to be dissipated, manifesting as more powerful storms, bigger deluges, deeper droughts. It was that simple. But a shadow of ignorance and denial had fallen over people who considered themselves children of the Enlightenment. It is for this reason that we now know this era as the Period of the Penumbra.

The loss of pet cats and dogs garnered particular attention among wealthy Westerners, but what was anomalous in 2023 soon became the new normal. . . . A shadow of ignorance and denial had fallen over people who considered themselves children of the Enlightenment.

It is clear that in the early twenty-first century, immediate steps should have been taken to begin a transition to a zero-net-carbon world. Staggeringly, the opposite occurred. At the very time that the urgent need for an energy transition became palpable, world production of greenhouse gases *increased*. This fact is so hard to understand that it calls for a closer look at what we know about this crucial juncture.

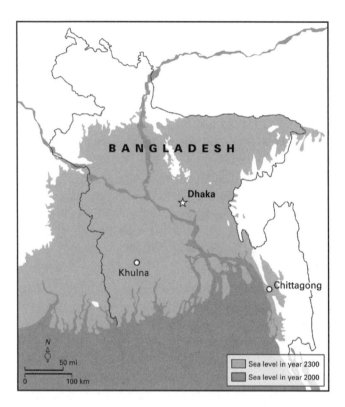

Bangladesh Among North Americans, Bangladesh—one of the poorest nations of the world—served as an ideological battleground. Self-described "Climate Hawks" used it to levy moral demands for greenhouse gas reductions so that it would not suffer inundation, while so-called "Climate Realists" insisted that only economic growth powered by cheap fossil fuels would make Bangladeshis wealthy enough to save themselves. In reality, "unfettered economic growth" made a handful of Bangladeshis wealthy enough to flee. The poor were left to the floods.

2

The Frenzy of Fossil Fuels

In the early Penumbral Period, physical scientists who spoke out about the potentially catastrophic effects of climate change were accused of being "alarmist" and of acting out of self-interest—to increase financial support for their enterprise, gain attention, or improve their social standing. At first, these accusations took the form of public denunciations; later they included threats, thefts, and the subpoena of private correspondence.[1] A crucial but under-studied incident was the legal seizing of notes from scientists who had documented the damage caused by a famous oil spill of the period, the 2011 British Petroleum Deepwater Horizon. Though leaders of the scientific community protested, scientists yielded to the demands, thus helping set the stage for further pressure on scientists from both governments and the industrial enterprises that governments subsidized and protected.[2] Then legislation was passed (particularly in the United States) that placed limits on what scientists could study and how they could study it, beginning with the notorious House Bill 819, better known as the "Sea Level Rise Denial Bill," passed in 2012 by the government of what was then the U.S. state of North Carolina (now part of the Atlantic Continental Shelf).[3] Meanwhile the Government Spending Accountability Act of 2012 restricted the ability of government scientists to attend conferences to share and analyze the results of their research.[4]

Though ridiculed when first introduced, the Sea Level Rise Denial Bill would become the model for the U.S. National Stability Protection Act of 2025, which led to the conviction and imprisonment of more than three hundred scientists for "endangering the safety and well-being of the general public with unduly alarming threats." By exaggerating the threat, it was argued, scientists were preventing the economic development essential for coping with climate change. When the scientists appealed, their convictions were upheld by the U.S. Supreme Court under the Clear and Present Danger doctrine, which permitted the government to limit speech deemed to represent an imminent threat.

Had scientists exaggerated the threat, inadvertently undermining the evidence that would later vindicate them? Certainly, narcissistic fulfillment played a role in the public positions that some scientists took, and in the early part of the Penumbral Period, funds flowed into climate research at the expense of other branches of science, not to mention other forms of intellectual and creative activity. Indeed, it is remarkable how little these extraordinarily wealthy nations spent to support artistic production; one explanation may be that artists were among the first to truly grasp the significance of the changes that were occurring. The most enduring literary work of this time is the celebrated science "fiction" trilogy by an American writer Kim Stanley Robinson—*Forty Signs of Rain*, *Fifty Degrees Below*, and *Sixty Days and Counting*.[5] Sculptor Dario Robleto also "spoke" to the issue, particularly species loss; his material productions have been lost, but responses to his work are recorded in contemporary accounts.[6] Some environmentalists also anticipated what was to come, notably the Australians Clive Hamilton and Paul Gilding. (Perhaps because Australia's population was highly educated and living on a continent at the edge of habitability, it was particularly sensitive to the changes under way.)[7] These "alarmists"—scientists and artists alike—were correct in their forecasts of an imminent shift in climate; in fact, by 2010 or so, it was clear that scientists had been underestimating the threat, as new developments outpaced early predictions of warming, sea level rise, and Arctic ice loss, among other parameters.[8]

It is difficult to understand why humans did not respond appropriately in the early Penumbral Period, when preventive measures were still possible. Many have sought an answer in the general phenomenon of *human adaptive optimism*, which later proved crucial for survivors. Even more elusive to scholars is why scientists, whose job it was to understand the threat and warn their societies—and who thought that they *did* understand the threat and that they *were* warning their societies—failed to appreciate the full magnitude of climate change.

To shed light on this question, some scholars have pointed to the epistemic structure of Western science, particularly in the late nineteenth and twentieth centuries, which was organized both intellectually and institutionally around "disciplines" in which specialists developed a high level of expertise in a small area of inquiry. This "reductionist" approach, sometimes credited to the seventeenth-century French philosophe René Descartes but not fully developed until the late nineteenth century, was believed to give intellectual power and vigor to investigations by focusing on singular elements of complex problems. "Tractability" was a guiding ideal of the time: problems that were too large or complex to be solved in their totality were divided into smaller, more manageable elements. While reductionism proved powerful in many domains, particularly quantum physics and medical diagnostics, it impeded investigations of complex systems. Reductionism also made it difficult for scientists to articulate the threat posed by climatic change, since many experts did not actually know very much about aspects of the problem beyond their expertise. (Other environmental problems faced similar challenges. For example, for years, scientists did not understand the role of polar stratospheric clouds in severe ozone depletion in the still-glaciated Antarctic region because "chemists" working on the chemical reactions did not even know that there *were* clouds in the polar stratosphere!) Even scientists who had a broad view of climate change often felt it would be inappropriate for them to articulate it, because that would require them to speak beyond their expertise, and seem to be taking credit for other people's work.

Responding to this, scientists and political leaders created the IPCC to bring together the diverse specialists needed to speak to the whole problem. Yet, perhaps because of the diversity of specialist views represented, perhaps because of pressures from governmental sponsors, or perhaps because of the constraints of scientific culture already mentioned,

> Reductionism also made it difficult for scientists to articulate the threat posed by climatic change, since many experts did not actually know very much about aspects of the problem beyond their expertise.

the IPCC had trouble speaking in a clear voice. Other scientists promoted the ideas of systems science, complexity science, and, most pertinent to our purposes here, earth systems science, but these so-called holistic approaches still focused almost entirely on natural systems, omitting from consideration the social components. Yet in many cases, the social components were the dominant system drivers. It was often said, for example, that climate change was caused by increased atmospheric concentrations of greenhouse gases. Scientists understood that those greenhouse gases were accumulating because of the activities of human beings—deforestation and fossil fuel combustion—yet they rarely said that the cause was *people*, and their patterns of conspicuous consumption.

Other scholars have looked to the roots of Western natural science in religious institutions. Just as religious orders of prior centuries had demonstrated moral rigor through extreme practices of asceticism in dress, lodging, behavior, and food—in essence, practices of physical self-denial—so, too, did physical scientists of the twentieth and twenty-first centuries attempt to demonstrate their intellectual rigor through practices of intellectual self-denial.[9] These practices led scientists to demand an excessively stringent standard for accepting claims of any kind, even those involving imminent threats. In an almost childlike attempt to demarcate their practices from those of older explanatory traditions, scientists felt it necessary to prove to themselves and the world how strict they were in their intellectual standards. Thus, they placed the burden of proof on novel claims—even empirical claims about phenomena that their theories predicted. This included claims about changes in the climate.

Some scientists in the early twenty-first century, for example, had recognized that hurricanes were intensifying.

This was consistent with the expectation—based on physical theory—that warmer sea surface temperatures in regions of cyclogenesis could, and likely would, drive either more hurricanes or more intense ones. However, they backed away from this conclusion under pressure from their scientific colleagues. Much of the argument surrounded the concept of *statistical significance.* Given what we now know about the dominance of nonlinear systems and the distribution of stochastic processes, the then-dominant notion of a 95 percent confidence limit is hard to fathom. Yet overwhelming evidence suggests that twentieth-century scientists believed that a claim could be accepted only if, by the standards of Fisherian statistics, the possibility that an observed event could have happened by chance was less than 1 in 20. Many phenomena whose causal mechanisms were physically, chemically, or biologically linked to warmer temperatures were dismissed as "unproven" because they did not adhere to this standard of demonstration. Historians have long argued about why this standard was accepted, given that it had neither epistemological nor substantive mathematical basis. We have come to understand the 95 percent confidence limit as a social convention rooted in scientists' desire to demonstrate their disciplinary severity.

Western scientists built an intellectual culture based on the premise that it was worse to fool oneself into believing in something that did not exist than not to believe in something that did. Scientists referred to these positions, respectively, as "type I" and "type II" errors, and established protocols designed to avoid type I errors at almost all costs. One scientist wrote, "A type I error is often considered to be more serious, and therefore more important to avoid, than a type II error." Another claimed that type II errors were not errors at all, just "missed opportunities."[10] So while the pattern of weather events was clearly changing, many scientists insisted that these events could not yet be attributed with certainty to anthropogenic climate change. Even as lay citizens began to accept this link, the scientists who studied it did not.[11] More important, political leaders came to believe that they had more time to act than they really did. The irony of these beliefs need not be dwelt on; scientists missed the most important opportunity in human history, and the costs that ensued were indeed nearly "all costs."

> Western scientists built an intellectual culture based on the premise that it was worse to fool oneself into believing in something that did not exist than not to believe in something that did.

By 2012, more than 365 billion tons of carbon had been emitted to the atmosphere from fossil fuel combustion and cement production. Another 180 were added from deforestation and other land use changes. Remarkably, more than half of these emissions occurred *after* the mid-1970s—that is, *after* scientists had built computer models demonstrating that greenhouse gases would cause warming. Emissions continued to accelerate even after the UNFCCC was established: between 1992 and 2012, total CO_2 emissions increased by 38 percent.[12] Some of this increase was understandable, as energy use grew in poor nations seeking to raise their standard of living. Less explicable is why, at the very moment when disruptive climate change was becoming apparent, wealthy nations dramatically increased their production of fossil fuels. The countries most involved in this enigma were two of the world's richest: the United States and Canada.

A turning point was 2005, when the U.S. Energy Policy Act exempted shale gas drilling from regulatory oversight under the Safe Drinking Water Act. This statute opened the floodgates (or, more precisely, the wellheads) to massive increases in shale gas production.[13] U.S. shale gas production at that time was less than 5 trillion cubic feet (Tcf, with "feet" an archaic imperial unit roughly equal to a third of a meter) per annum. By 2035, it had increased to 13.6 Tcf. As the United States expanded shale gas production and exported the relevant technology, other nations followed. By 2035, total gas production had exceeded 250 Tcf per annum.[14]

This bullish approach to shale gas production penetrated Canada as well, as investor-owned companies raced to develop additional fossil fuel resources; "frenzy" is not too strong a word to describe the surge of activity that occurred. In the late twentieth century, Canada was considered an advanced nation with a high level of environmental sensitivity, but this changed around the year 2000 when Canada's government began to push for development of huge tar sand deposits in the province of Alberta, as well as shale gas in various parts of the country. The tar sand deposits (which the government preferred to call oil sands, because liquid oil had a better popular image than sticky tar) had been mined intermittently since the 1960s, but the rising cost of conventional oil now made sustained exploitation economically feasible. The fact that 70 percent of the world's known reserves were in Canada explains the government's reversed position on climate change: in 2011, Canada withdrew from the Kyoto Protocol to the UNFCCC.[15] Under the protocol, Canada had committed to cut its emissions by 6 percent, but its actual emissions instead increased more than 30 percent.[16]

> Canada was considered an advanced nation with a high level of environmental sensitivity, but this changed around the year 2000 when Canada's government began to push for development of huge tar sand deposits in the province of Alberta, as well as shale gas in various parts of the country.

Meanwhile, following the lead of the United States, the government began aggressively to promote the extraction of shale gas, deposits of which occurred throughout Canada. Besides driving up direct emissions of both CO_2 and CH_4 to the atmosphere (since many shale gas fields also contained CO_2, and virtually all wells leaked), the resulting massive increase in supply of natural gas led to a collapse in the market price, driving out nascent renewable energy industries everywhere except China, where government subsidies and protection for fledgling industries enabled the renewable sector to flourish.

Cheap natural gas also further undermined the already ailing nuclear power industry, particularly in the United States. To make matters worse, the United States implemented laws forbidding the use of biodiesel fuels—first by the military, and then by the general public—undercutting that emerging market as well.[17] Bills were passed on both the state and federal level to restrict the development and use of other forms of renewable energy—particularly in the highly regulated electricity generation industry—and to inhibit the sale of electric cars, maintaining the lock that fossil fuel companies had on energy production and use.[18]

Meanwhile, Arctic sea ice melted, and seaways opened that permitted further exploitation of oil and gas reserves in the north polar region. Again, scientists noted what was happening. By the mid-2010s, the Arctic summer sea had lost about 30 percent of its areal extent compared to 1979, when high-precision satellite measurements were first made; the average loss was rather precisely measured at 13.7 percent per decade from 1979 to 2013.[19] When the areal extent of summer sea ice was compared to earlier periods using additional data from ships, buoys, and airplanes, the total summer loss was nearly 50 percent. The year 2007 was particularly worrisome, as the famous

Northwest Passage—long sought by Arctic explorers—opened, and the polar seas became fully navigable for the first time in recorded history. Scientists understood that it was only a matter of time before the Arctic summer would be ice-free, and that this was a matter of grave concern. But in business and economic circles it was viewed as creating opportunities for further oil and gas exploitation.[20] One might have thought that governments would have stepped in to prevent this ominous development—which could only exacerbate climate change—but governments proved complicit. One example: in 2012 the Russian government signed an agreement with American oil giant ExxonMobil, allowing the latter to explore for oil in the Russian Arctic in exchange for Russian access to American shale oil drilling technology.[21]

How did these wealthy nations—rich in the resources that would have enabled an orderly transition to a zero-net-carbon infrastructure—justify the deadly expansion of fossil fuel production? Certainly, they fostered the growing denial that obscured the link between climate change and fossil fuel production and consumption. They also entertained a second delusion: that natural gas from shale could offer a "bridge to renewables." Believing that conventional oil and gas resources were running out (which they were, but at a rate insufficient to avoid disruptive climate change), and stressing that natural gas produced only half as much CO_2 as coal, political and economic leaders—and even many climate scientists and "environmentalists"—persuaded themselves and their constituents that promoting shale gas was an environmentally and ethically sound approach.

This line of reasoning, however, neglected several factors. First, *fugitive emissions*—CO_2 and CH_4 that escaped from wellheads into the atmosphere—greatly accelerated warming. (As with so many climate-related phenomena, scientists had foreseen this, but their predictions were buried in specialized journals.) Second, most analyses of the greenhouse gas benefits of gas were based on the assumption that it would replace coal in electricity generation where the benefits, if variable, were nevertheless fairly clear. However, as gas became cheap, it came to be used increasingly in transportation and home heating, where losses in the distribution system negated many of the gains achieved in electricity generation. Third, the calculated benefits were based on the assumption that gas would replace coal, which it did in some regions (particularly in the United States and some parts of Europe), but elsewhere (for example Canada) it mostly replaced nuclear and hydropower. In many regions cheap gas simply became an additional energy source, satisfying expanding demand without replacing other forms of fossil fuel energy production. As new gas-generating power plants were built, infrastructures based on fossil fuels were further locked in, and total global emissions continued to rise. The argument for the climatic benefits of natural gas presupposed that net CO_2 emissions would fall, which would have required strict restrictions on coal and petroleum use in the short run and the phase-out of gas as well in the long run.[22] Fourth, the analyses mostly omitted the cooling effects of aerosols from coal, which although bad for human health had played a significant role in keeping warming below the level it would otherwise have already reached. Fifth, and perhaps most important, the sustained low prices of fossil fuels, supported by continued subsidies and a lack of external cost accounting, undercut efficiency efforts and weakened emerging markets for solar, wind, and biofuels (including crucial liquid biofuels for aviation).[23] Thus, the bridge to a zero-carbon future collapsed before the world had crossed it.

The net result? Fossil fuel production escalated, greenhouse gas emissions increased, and climate disruption accelerated. In 2001, the IPCC had predicted that atmospheric CO_2 would double by 2050.[24] In fact, that benchmark was met by 2042. Scientists had expected a mean global warming of 2 to 3 degrees Celsius; the actual figure was 3.9 degrees. Though originally merely a benchmark for discussion with no particular physical meaning, the doubling of CO_2 emissions turned out to be quite significant: once the corresponding temperature rise reached 4 degrees, rapid changes began to ensue.

By 2040, heat waves and droughts were the norm. Control measures—such as water and food rationing and Malthusian "one-child" policies—were widely implemented. In wealthy countries, the most hurricane- and tornado-prone regions were gradually but steadily depopulated, putting increased social pressure on areas less subject to those hazards. In poor nations, conditions were predictably worse: rural portions of Africa and Asia began experiencing significant depopulation from out-migration, malnutrition-induced disease and infertility, and starvation. Still, sea level had risen only 9 to 15 centimeters around the globe, and coastal populations were mainly intact.

Then, in the Northern Hemisphere summer of 2041, unprecedented heat waves scorched the planet, destroying food crops around the globe. Panic ensued, with food riots in virtually every major city. Mass migration of undernourished and dehydrated individuals, coupled with explosive increases in insect populations, led to widespread outbreaks of typhus, cholera, dengue fever, yellow fever, and viral and retroviral agents never before seen. Surging insect populations also destroyed huge swaths of forests in Canada, Indonesia, and Brazil. As social order began to break down in the 2050s, governments were overthrown, particularly in Africa, but also in many parts of Asia and Europe, further decreasing social capacity to deal with increasingly desperate populations. As the Great North American Desert surged north and east, consuming the High Plains and destroying some of the world's most productive farmland, the U.S. government declared martial law to prevent food riots and looting. A few years later, the United States announced plans with Canada for the two nations to begin negotiations toward the creation of the United States of North America, to develop an orderly plan for resource-sharing and northward population relocation. The European Union announced similar plans for voluntary northward relocation of eligible citizens from its southernmost regions to Scandinavia and the United Kingdom.

While governments were straining to maintain order and provide for their people, leaders in Switzerland and India—two countries that were rapidly losing substantial portions of their glacially-sourced water resources—convened the First International Emergency Summit on Climate Change, organized under the rubric of Unified Nations for Climate Protection (the former United Nations having been discredited and disbanded over the failure of the UNFCCC). Political, business, and religious leaders met in Geneva and Chandigarh to discuss emergency action. Many said that the time had come to make the switch to zero-carbon energy sources. Others argued that the world could not wait the ten to fifty years required to alter the global energy infrastructure, much less the one hundred years it would take for atmospheric CO_2 to diminish. In response, participants hastily wrote and signed the

Mass migration of undernourished and dehydrated individuals, coupled with explosive increases in insect populations, led to widespread outbreaks of typhus, cholera, dengue fever, yellow fever, and viral and retroviral agents never before seen.

Unified Nations Convention on Climate Engineering and Protection (UNCCEP), and began preparing blueprints for the International Climate Cooling Engineering Project (ICCEP).

As a first step, ICCEP launched the International Aerosol Injection Climate Engineering Project (IAICEP, pronounced ay-yi-yi-sep) in 2052.[25] Sometimes called the Crutzen Project after the scientist who first suggested the idea in 2006, projects like this engendered heated public opposition when first proposed in the early twenty-first century but had widespread support by mid-century— from wealthy nations anxious to preserve some semblance of order, from poor nations desperate to see the world do something to address their plight, and from frantic low-lying Pacific Island nations at risk of being submerged by rising sea levels.[26]

IAICEP began to inject submicrometer-size sulfate particles into the stratosphere at a rate of approximately 2.0 teragrams per year, expecting to reduce mean global temperature by 0.1 degrees Celsius annually from 2059 to 2079. (In the meantime, a substantial infrastructural conversion to renewable energy could have been achieved.) Initial results were encouraging: during the first three years of implementation, temperature decreased as expected and the phase-out of fossil fuel production commenced. However, in the project's fourth year, an anticipated but discounted side effect occurred: the shutdown of the Indian monsoon. (By decreasing incoming solar radiation, IAICEP also decreased evaporation over the Indian Ocean, and hence the negative impact on the monsoon.) As crop failures and famine swept across India, one of IAICEP's most aggressive promoters now called for its immediate cessation.

IAICEP was halted in 2063, but a fatal chain of events had already been set in motion. It began with *termination shock*—that is, the abrupt increase in global temperatures following the sudden cessation of the project. Once again, this phenomenon had been predicted, but IAICEP advocates had successfully argued that, given the emergency conditions, the world had no choice but to take the risk.[27] In the following eighteen months, temperature rapidly rebounded, regaining not just the 0.4 degrees Celsius that had been reduced during the project but an additional 0.6 degrees. This rebound effect pushed the mean global temperature increase to nearly 5 degrees Celsius.

Whether it was caused by this sudden additional heating or was already imminent is not known, but the greenhouse effect then reached a global tipping point. By 2060, Arctic summer ice was completely gone. Scores of species perished, including the iconic polar bear—the Dodo bird of the twenty-first century. While the world focused on these highly visible losses, warming had meanwhile accelerated the less visible but widespread thawing of Arctic permafrost. Scientists monitoring the phenomenon observed a sudden increase in permafrost thaw and CH_4 release. Exact figures are not available, but the estimated total carbon release of Arctic CH_4 during the next decade may have reached over 1,000 gigatonnes, effectively doubling the total atmospheric carbon load.[28] This massive addition of carbon led to what is known as the Sagan effect (sometimes more dramatically called the Venusian death): a strong positive feedback loop between warming and CH_4 release. Planetary temperature increased by an additional 6 degrees Celsius over the 5-degree rise that had already occurred.

The ultimate blow for Western civilization came in a development that, like so many others, had long been discussed but rarely fully assimilated as a realistic threat: the collapse of the West Antarctica Ice Sheet. Technically, what happened in West Antarctica was not a collapse; the ice sheet did not fall in on itself, and it did not happen all at once. It was more of a rapid disintegration. Post hoc failure analysis shows that extreme heat in the Northern Hemisphere disrupted normal patterns of ocean circulation, sending exceptionally warm surface waters into the southern ocean that destabilized the ice sheet from below. As large pieces of ice shelf began to separate from the main ice sheet, removing the bulwark that had kept the sheet on the Antarctic Peninsula, sea level began to rise rapidly.

Social disruption hampered scientific data-gathering, but some dedicated individuals—realizing the damage could not be stopped—sought, at least, to chronicle it. Over the course of the next two decades (from 2073 to 2093), approximately 90 percent of the ice sheet broke apart, disintegrated, and melted, driving up sea level approximately five meters across most of the globe. Meanwhile, the Greenland Ice Sheet, long thought to be less stable than the Antarctic Ice Sheet, began its own disintegration. As summer melting reached the center of the Greenland Ice Sheet, the east side began to separate from the west. Massive ice breakup ensued, adding another two meters to mean global sea level rise.[29] These cryogenic events were soon referred to as the Great Collapse, although some scholars now use the term more broadly to include the interconnected social, economic, political, and demographic collapse that ensued.

Analysts had predicted that an eight-meter sea level rise would dislocate 10 percent of the global population. Alas, their estimates proved low: the reality was closer to 20 percent. Although records for this period are incomplete, it is likely that during the Mass Migration 1.5 billion people were displaced around the globe, either directly from the impacts of sea level rise or indirectly from other impacts of climate change, including the secondary dislocation of inland peoples whose towns and villages were overrun by eustatic refugees. Dislocation contributed to the Second Black Death, as a new strain

These cryogenic events were soon referred to as the Great Collapse, although some scholars now use the term more broadly to include the interconnected social, economic, political, and demographic collapse that ensued.

of the bacterium *Yersinia pestis* emerged in Europe and spread to Asia and North America. In the Middle Ages, the Black Death killed as much as half the population of some parts of Europe; this second Black Death had similar effects.[30] Disease also spread among stressed nonhuman populations. Although accurate statistics are scant because twentieth-century scientists did not have an inventory of total global species, it is not unrealistic to estimate that 60 to 70 percent of species were driven to extinction. (Five previous mass extinctions were known to scientists of the Penumbral Period, each of which was correlated to rapid greenhouse gas level changes, and each of which destroyed more than 60 percent of identifiable species—the worst reached 95 percent. Thus, 60–70 percent is a conservative estimate insofar as most of these earlier mass extinctions happened more slowly than the anthropogenic mass extinction of the late Penumbral Period.)[31]

> Dislocation contributed to the Second Black Death, as a new strain of the bacterium *Yersinia pestis* emerged in Europe and spread to Asia and North America. In the Middle Ages, the Black Death killed as much as half the population of Europe; this second Black Death had similar effects.

There is no need to rehearse the details of the human tragedy that occurred; every schoolchild knows of the terrible suffering. Suffice it to say that total losses—social, cultural, economic, and demographic—were greater than any in recorded human history. Survivors' accounts make clear that many thought the end of the human race was near. Indeed, had the Sagan effect continued, warming would not have stopped at 11 degrees, and a runaway greenhouse effect would have followed.

However, around 2090 (the date cannot be determined from extant records), something occurred whose exact character remains in dispute. Japanese genetic engineer Akari Ishikawa developed a form of lichenized fungus in which the photosynthetic partner consumed atmospheric CO_2 much more efficiently than existing forms, and was able to grow in a wide diversity of environmental conditions. This pitch-black lichen, dubbed *Pannaria ishikawa*, was deliberately released from Ishikawa's laboratory, spreading rapidly throughout Japan and then across most of the globe. Within two decades, it had visibly altered the visual landscape and measurably altered atmospheric CO_2, starting the globe on the road to atmospheric recovery and the world on the road to social, political, and economic recovery.

In public pronouncements, the Japanese government has maintained that Ishikawa acted alone, and cast her as a criminal renegade. Yet many Japanese citizens have seen her as a hero, who did what their government could not, or would not, do. Most Chinese scholars reject both positions, contending that the Japanese government, having struggled and failed to reduce Japan's own carbon emissions, provided Ishikawa with the necessary resources and then turned a blind eye toward its dangerous and uncertain character. Others blame (or credit) the United States, Russia, India, or Brazil, as well as an international consortium of financiers based in Zurich. Whatever the truth of this matter, Ishikawa's actions slowed the increase of atmospheric CO_2 dramatically.

Humanity was also fortunate in that a so-called "Grand Solar Minimum" reduced incoming solar radiation during the twenty-second century by 0.5%, offsetting some of the excess CO_2 that had accumulated, and slowing the rise of surface and oceanic temperatures for nearly a century, during which time survivors in northern inland regions of Europe, Asia, and North America, as well as inland and high-altitude regions of South America, were able to begin to regroup and rebuild. The human populations of Australia and Africa, of course, were wiped out.

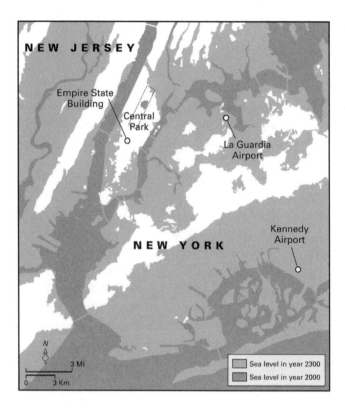

New York City in the twenty-fourth century Once the financial capital of the world, New York began in the early twenty-first century to attempt to defend its elaborate and expensive infrastructure against the sea. But that infrastructure had been designed and built with an expectation of constant seas and was not easily adapted to continuous, rapid rise. Like the Netherlands, New York City gradually lost its struggle. Ultimately, it proved less expensive to retreat to higher ground, abandoning centuries' worth of capital investments.

3

Market Failure

To the historian studying this tragic period of human history, the most astounding fact is that the victims *knew what was happening and why.* Indeed, they chronicled it in detail precisely *because* they knew that fossil fuel combustion was to blame. Historical analysis also shows that Western civilization had the technological know-how and capability to effect an orderly transition to renewable energy, yet the available technologies were not implemented in time.[1] As with all great historical events, there is no easy answer to the question of why this catastrophe occurred, but key factors stand out. The thesis of this analysis is that Western civilization became trapped in the grip of two inhibiting ideologies: *positivism* and *market fundamentalism.*

Twentieth-century scientists saw themselves as the descendants of an empirical tradition often referred to as *positivism*—after the nineteenth-century French philosopher, Auguste Comte, who developed the concept of "positive" knowledge (as in, "absolutely, positively true")—but the overall philosophy is more accurately known as *Baconianism*. This philosophy held that through experience, observation, and experiment, one could gather reliable knowledge about the natural world, and that this knowledge would empower its holder. Experience justified the first part of the philosophy (we have recounted how twentieth-century scientists anticipated the consequences of climate change), but the second part—that this knowledge would translate into power—proved less accurate. Although billions of dollars were spent on climate research in the late twentieth and early twenty-first centuries, the resulting knowledge had little impact on the crucial economic and technological policies that drove the continued use of fossil fuels.

A key attribute of the period was that power did not reside in the hands of those who understood the climate system, but rather in political, economic, and social institutions that had a strong interest in maintaining the use of fossil fuels. Historians have labeled this system the *carbon-combustion complex*: a network of powerful industries comprising fossil fuel producers, industries that served energy companies (such as drilling and oil field service companies and large construction firms), manufacturers whose products relied on inexpensive energy (especially automobiles and aviation, but also aluminum and other forms of smelting and mineral processing), financial institutions that serviced their capital demands, and advertising, public relations, and marketing firms

who promoted their products. Maintaining the carbon-
combustion complex was clearly in the self-interest of
these groups, so they cloaked this fact behind a network
of "think tanks" that issued challenges to scientific knowledge they found threatening.[2] Newspapers often quoted
think tank employees as if they were climate researchers,
juxtaposing their views against those of epistemologically
independent university or government scientists. This
practice gave the public the impression that the science
was still uncertain, thus undermining the sense that it was
time to act.[3] Meanwhile, scientists continued to do science, believing, on the one hand, that it was inappropriate
for them to speak on political questions (or to speak in
the emotional register required to convey urgency) and,
on the other hand, that if they produced abundant and
compelling scientific information (and explained it calmly
and clearly), the world would take steps to avert disaster.

Many scientists, to their credit, recognized the difficulties they were facing, and grappled with how to communicate their knowledge effectively.[4] Some tried to create
institutional structures to support less reductionist modes
of inquiry that analyzed broad patterns and the interactions between natural and social systems. While they were
making some headway, a large part of Western society was
rejecting that knowledge in favor of an empirically inad-
equate yet powerful ideological system. Even at the time,
some recognized this system as a quasi-religious faith,
hence the label *market fundamentalism*.

Market fundamentalism—and its various strands and interpretations known as free market fundamentalism, neoliberalism, laissez-faire economics, and laissez-faire capitalism—was a two-pronged ideological system. The first prong held that societal needs were served most efficiently in a free market economic system. Guided by the "invisible hand" of the marketplace, individuals would freely respond to each other's needs, establishing a net balance between solutions ("supply") and needs ("demand"). The second prong of the philosophy maintained that free markets were not merely a good or even the best manner of satisfying material wants: they were the *only* manner of doing so that did not threaten personal freedom.

The crux of this second point was the belief that marketplaces represented distributed power. Various individuals making free choices held power in their hands, preventing its undue concentration in centralized government. Conversely, centrally planned economies entailed not just the concentration of economic power, but of political power as well. Thus, to protect personal liberty—political, civic, religious, artistic—economic liberty had to be preserved. This aspect of the philosophy, called *neoliberalism*, hearkened back to the liberalism of the eighteenth- and nineteenth-century Enlightenment, particularly the works of Adam Smith, David Hume, John Locke, and, later, John Stuart Mill. Reacting to the dominant form of Western governance in their time—that is, monarchy—these thinkers lionized personal liberty in contrast to control by unjust, arbitrary, and often incompetent despots. At a time when some political leaders were imagining alternatives to despotic monarchy, many viewed the elevation of individual rights as a necessary response. In the late eighteenth century, these views influenced the architects of the American Republic and the early, "liberal" phase of the French Revolution. Even then, however, such views were more idealistic than realistic. The late-eighteenth-century U.S. Constitution preserved and validated race-based slavery; even when that nation abolished slavery in the mid-nineteenth century, it preserved economic and social apartheid for more than a century thereafter. In Europe, the French Revolution collapsed in a wave of violence and the restoration of autocratic rule under Napoleon Bonaparte.

In the nineteenth century, power became concentrated in the hands of industrialists (the "robber barons," monopolies, and "trusts" of the United States and elsewhere), challenging liberal conceptions of the desirability of weak political governance. In Europe, the German philosopher Karl Marx argued that an inherent feature of the capitalist system was the concentration of wealth and power in a ruling class that siphoned off the surplus value produced by workers. Industrialists not only employed workers under brutal and tyrannical conditions (the nineteenth-century "satanic mills"), they also corrupted democratic processes through bribery and extortion and distorted the marketplace through a variety of practices. A powerful example is the development and expansion of American railroads. Supply of these "roads to nowhere" was heavily subsidized, and the demand for them was manufactured at the expense of displaced native peoples and natural environment of the American West.[5]

Marx's analysis inspired popular leaders in many nation-states then in existence—for example, Russia, China, Vietnam, Ghana, and Cuba—to turn to communism as an alternative economic and social system. Meanwhile, the capitalist United States abolished slavery and made adjustments to remedy power imbalances and losses of liberty due to the concentration of wealth. Among other reforms, the federal government introduced *antitrust laws* to prevent monopolistic practices, established worker protections such as limits on the length of the working day and prohibitions on child labor, and developed a progressive income tax. By the early twentieth century, it was clear that capitalism in its pure, theoretical form did not exist, and few could argue for its desirability: the failures were too obvious. Intellectuals came to see the invisible hand, akin to the hand of God, as the quasi-religious notion that it was. The Great Depression of the 1930s—from which Europe and the United States emerged only through the centralized mobilization of World War II—led scholars and political leaders to view the idea of self-regulating markets as a myth. After WWII, most non-communist states became "mixed" economies with a large degree of both individual and corporate freedom and significant government involvement in markets, including extensive systems of taxes, tariffs, subsidies, regulation of banks and exchanges, and immigration control.[6]

Meanwhile communism, which had spread throughout Eurasia and to some parts of Africa and Latin and South America, was revealing even worse failures than capitalism. Communist economies proved grossly inefficient at delivering goods and services; politically, early ideas of mass empowerment gave way to tyrannical and brutal dictatorship. In the Soviet Union under Joseph Stalin, tens of millions died in purges, forced collectivization of agriculture, and other forms of internal violence. Tens of millions died in China as well during the "Great Leap Forward"—the attempt by 毛泽东 (Mao Zedong) to force rapid industrialization—and many more in the so-called Cultural Revolution of the First People's Republic of China (PRC).[7]

Following World War II, the specter of Russian communism's spread into Eastern (and possibly even Western) Europe—thus affecting U.S. access to markets and stoking fears that the West could sink back into economic depression—led the United States to take a strong position against Soviet expansion. Conversely, the Soviet Union's desire to control its western flanks in light of historic vulnerability led to the political occupation and control of Eastern Europe. The resulting Cold War (1945–1989) fostered a harshly polarized view of economic systems, with "communists" decrying the corruption of the capitalist system and "capitalists" condemning the tyranny and violence in communist regimes (acknowledging here that in practice neither system was purely communist nor purely capitalist). Perhaps because of the horrible violence in the East, many Western intellectuals came to see everything associated with communism as evil, even— and crucially for our story—modest or necessary forms of intervention in the marketplace such as progressive taxation and environmental regulation, and humanitarian interventions such as effective and affordable regimes of health care and birth control.

Neoliberalism was developed by a group of thinkers— most notably Austrian Friedrich von Hayek and American Milton Friedman—who were particularly sensitive to the issue of repressive centralized government. In two key works, von Hayek's *Road to Serfdom* and Friedman's *Capitalism and Freedom*, they developed the crucial "neo-" component of neoliberalism: the idea that free market systems were the only economic systems that did not threaten individual liberty.

Neoliberalism was initially a minority view. In the 1950s and 1960s, the West experienced high overall prosperity, and individual nations developed mixed economies that suited their own national cultures and contexts. Things began to shift in the late 1970s and 1980s, when Western economies stalled and neoliberal ideas attracted world leaders searching for answers to their countries' declining economic performance, such as Margaret Thatcher in the United Kingdom and Ronald Reagan in the United States.

Friedman became an advisor to President Reagan; in 1991, von Hayek received the Presidential Medal of Freedom from President George H. W. Bush.[8]

An ironic aspect of this element of our story is that Friedrich von Hayek had great respect and admiration for the scientific enterprise, seeing it as the historical companion to enterprise capitalism. By fostering commerce, von Hayek suggested, science and industry were closely linked to the rise of capitalism and the growth of political freedom; this view was shared by mid-twentieth-century advocates for an expanded role of government in promoting scientific investigations.[9] However, when environmental science showed that government action was needed to protect citizens and the natural environment from unintended harms, the carbon-combustion complex began to treat science as an enemy to be fought by whatever means necessary. The very science that had led to U.S. victory in World War II and dominance in the Cold War became the target of skepticism, scrutiny, and attack. (Science was also attacked in communist nations for the same basic reason—it came into conflict with political ideology.)[10] The end of the Cold War (1989–1991) was a source of celebration for citizens who had lived under the yoke of oppressive Soviet-style governance; it also ignited a slow process of overdue reforms in the First People's Republic of China. But for many observers in the West, the Soviet Union's collapse gave rise to an uncritical triumphalism, proof of the absolute superiority of the capitalist system.

Some went further, arguing that if capitalism was a superior system, then the best form of capitalism lay in its purest form. While it is possible that some academic economists and intellectuals genuinely held this view, it was the industrialists and financiers who perceived large opportunities in less regulated markets who did the most to spread and promote it. As a result, the 1990s and 2000s featured a wave of deregulation that weakened consumer, worker, and environmental protections. A second Gilded Age reproduced concentrations of power and capital not seen since the nineteenth century, with some of the accumulated capital used to finance think tanks that further promoted neoliberal views. (And some of the rest reinvested in fossil fuel production.) Most important for our purposes, neoliberal thinking led to a refusal to admit the most important limit of capitalism: market failure.

When scientists discovered the limits of planetary sinks, they also discovered market failure. The toxic effects of DDT, acid rain, the depletion of the ozone layer, and climate change were serious problems for which

> When environmental science showed that government action was needed to protect citizens and the natural environment from unintended harms, the carbon-combustion complex began to treat science as an enemy to be fought by whatever means necessary. The very science that had led to U.S. victory in World War II and dominance in the Cold War became the target of skepticism, scrutiny, and attack.

markets did not provide a spontaneous remedy. Rather, government intervention was required: to raise the market price of harmful products, to prohibit those products, or to finance the development of their replacements. But because neoliberals were so hostile to centralized government, they had, as Americans used to say, "painted themselves into a corner." The American people had been

> The toxic effects of DDT, acid rain, the depletion of the ozone layer, and climate change were serious problems for which markets did not provide a spontaneous remedy.

persuaded, in the words of U.S. President Ronald Reagan (r. 1980–1988), that government was "the problem, not the solution." Citizens slid into passive denial, accepting the contrarian arguments that the science was unsettled. Lacking widespread support, government leaders were unable to shift the world economy to a net carbon-neutral energy base. As the world climate began to spin out of control and the implications for market failure became indisputable, scientists came under attack, blamed for problems they had not caused, but had documented.

Physical scientists were chief among the individuals and groups who tried to warn the world of climate change, both before and as it happened. (In recognition of what they tried to achieve, millions of survivors have taken their names as middle names.) Artists noted the tendency to ignore warning signs, such as the mid-twentieth-century Canadian songwriter Leonard Cohen, who sang "We asked for signs. The signs were sent." Social scientists introduced the concept of "late lessons from early warnings" to describe a growing tendency to neglect information. As a remedy, they promoted a *precautionary principle*, whereby early action would prevent later damage. The *precautionary principle* was a formal instantiation of what had previously been thought of as common sense, reflected in the nineteenth-century European and American adages, "A stitch in time saves nine" and "An ounce of prevention is worth a pound of cure." Yet this traditional wisdom was swept away in neoliberal hostility toward planning and an overconfident belief in the power of markets to respond to social problems as they arose. (Indeed, neoliberals believed markets so powerful they could "price in" futures that had not happened yet—pre-solving problems as it were, a brilliant case of wishful fantasy that obviated the need for hateful planning.) Another of the many ironies of the Penumbral Period is that the discipline of economics—rooted in the ancient Greek concept of household management (*oikos*, "house," and *nomos*, "laws" or "rules")—failed to speak to the imperative of a managed transition to a new energy system. The idea of managing energy use and

> The *precautionary principle* was a formal instantiation of what had previously been thought of as common sense, reflected in the nineteenth-century European and American adages, "A stitch in time saves nine" and "An ounce of prevention is worth a pound of cure."

controlling greenhouse gas emissions was anathema to the neoliberal economists whose thinking dominated at this crucial juncture. Thus, no planning was done, no precautions were taken, and the only management that finally ensued was disaster management.

Discerning neoliberals acknowledged that the free market was not really free; interventions were everywhere. Some advocated eliminating subsidies for fossil fuels and creating "carbon" markets.[11] Others recognized that certain interventions could be justified. Von Hayek himself was not opposed to government intervention per se; indeed, as early as 1944, he rejected the term *laissez-faire* as misleading because he recognized legitimate realms of government intervention: "The successful use of competition as the principle of social organization precludes certain types of coercive interference with economic life, but it admits of . . . and even requires [others]," he wrote. In his view, legitimate interventions included paying for signposts on roads, preventing "harmful effects of deforestation, of some methods of farming, or of the noise and smoke of factories," prohibiting the use of poisonous substances, limiting working hours, enforcing sanitary conditions in workplaces, controlling weights and measures, and preventing violent strikes.[12] Von Hayek simply (and reasonably) believed that if the government were to carry out such functions, and particularly if doing so *selectively* limited the freedom of particular groups or individuals, then the justification for the intervention should be clear and compelling. Given the events recounted here, it is hard to imagine why anyone in the twentieth century would have argued against government protection of the natural environment on which human life depends. Yet such arguments were not just made, they dominated the public sphere.

The ultimate paradox was that neoliberalism, meant to ensure individual freedom above all, led eventually to a situation that necessitated large-scale government intervention. Classical liberalism was centered on the idea of individual liberty, and in the eighteenth century most individuals had precious little of it—economic or otherwise. But by the mid-twentieth century this situation had changed dramatically: slavery was formally outlawed in the nineteenth century, and monarchies and other forms of empire were increasingly replaced by various forms of "liberal" democracy. In the West, individual freedoms—both formal and informal—probably peaked around the time von Hayek was writing, or shortly thereafter. By the end of the twentieth century, Western citizens still held the formal rights of voting, various forms of free thought and expression, and freedom of employment and travel. But actionable freedom was decreasing, first as economic power was increasingly concentrated in a tiny elite, who came to be known as the "1 percent," and then in a political elite propelled to power as the climate crisis forced dramatic interventions to relocate citizens displaced by sea level rise and desertification, to contain contagion, and to prevent mass famine. And so the development that the neoliberals most dreaded—centralized government and loss of personal choice—was rendered essential by the very policies that they had put in place.

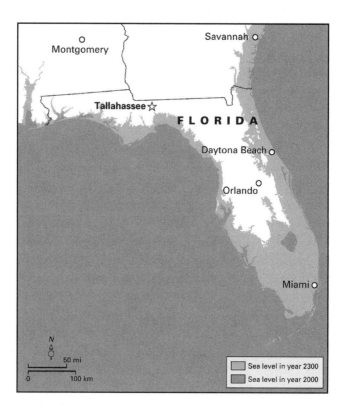

| | Sea level in year 2300 |
| | Sea level in year 2000 |

The former state of Florida (part of the former United States) In one of the many paradoxes of history, the inhabitants of late-twentieth-century Florida were engaged in a grand project to save an enormous sea-level wetlands region known as the Everglades from urban growth and the diversion of freshwater to urban and agricultural use. Yet even the low-end estimates of twenty-first century sea-level rise rendered this effort pointless due to inundation; what actually happened cost Floridians both the Everglades and many of their major cities.

Epilogue

As the devastating effects of the Great Collapse began to appear, the nation-states with democratic governments—both parliamentary and republican—were at first unwilling and then unable to deal with the unfolding crisis. As food shortages and disease outbreaks spread and sea level rose, these governments found themselves without the infrastructure and organizational ability to quarantine and relocate people.

In China, the situation was somewhat different. Like other post-communist nations, China had taken steps toward liberalization but still retained a powerful centralized government. When sea level rise began to threaten coastal areas, China rapidly built new inland cities and villages and relocated more than 250 million people to higher, safer ground.[1] The relocation was not easy; many older citizens, as well as infants and young children, could not manage the transition. Nonetheless, survival rates exceeded 80 percent. To many survivors—in what might be viewed as a final irony of our story—China's ability to weather disastrous climate change vindicated the necessity of centralized government, leading to the establishment of the Second People's Republic of China (SPRC) (also sometimes referred to as Neocommunist China) and inspiring similar structures in other, reformulated nations. By blocking anticipatory action, neoliberals did more than expose the tragic flaws in their own system: they fostered expansion of the forms of governance they most abhorred.

Today, we remain engaged in a vigorous intellectual discussion of whether, now that the climate system has finally stabilized, decentralization and redemocratization may be considered. Many academics, in the spirit of history's great thinkers, hope that such matters may be freely debated. Others consider that outcome wishful, in light of the dreadful events of the past, and reject the reappraisal that we wish to invite here. Evidently, the Penumbra falls even today—and likely will continue to fall for years, decades, and perhaps even centuries to come.

Lexicon of Archaic Terms

Anthropocene The geological period, beginning in approximately 1750 with the start of the Industrial Revolution, when humans have become geological agents whose activities effectively compete with, and begin to overwhelm, geophysical, geochemical, and biological processes.

Baconianism A philosophy, generally attributed to the English jurist Sir Francis Bacon (1561–1626), that held that through experience, observation, and experiment, one could gather reliable knowledge about the natural world and this knowledge would empower its holder. The fallacy of Baconianism was clearly demonstrated by the powerlessness of scientists, in the late twentieth and twenty-first centuries, to effect meaningful action on climate change despite their acute knowledge of it.

bridge to renewables The logical fallacy, popular in the first decades of the twenty-first century, that the problem of greenhouse gas emissions from fossil fuel combustion could be solved by burning more fossil fuels, particularly natural gas. The fallacy rested on an incomplete analysis, which considered only physical by-products of combustion, particularly in electricity generation, and not the other factors that controlled overall energy use and net release of greenhouse gases to the atmosphere.

capitalism A form of socioeconomic organization that dominated Western Europe and North America from the sixteenth to the twentieth centuries, in which the means of production and distribution of goods and services were owned either by individuals or by government-chartered legal entities called "corporations." Typically these entities were operated for-profit, with the surplus value produced by workers funneled to owners, managers, and "investors," third parties who owned "stock" in a company but had liability neither for its debts nor its social consequences. The separation of work from ownership produced a concentration of wealth amongst a tiny elite, who could then purchase more favorable laws and regulations from their host governments. One popular notion about capitalism of the period was that it operated through a process of creative destruction. Ultimately, capitalism was paralyzed in the face of the rapid climate destabilization it drove, destroying itself.

carbon-combustion complex The interlinked fossil fuel extraction, refinement, and combustion industries, financiers, and government "regulatory" agencies that enabled and defended destabilization of the world's climate in the name of employment, growth, and prosperity.

communism A political ideology rooted in the early European industrial revolution that favored workers' rights over those of financiers and managers, and state planning over market forces. Adopted primarily in Eurasia in the early to mid-twentieth century, it was largely abandoned in the late twentieth century due to its failures to deliver on its economic promises, and its susceptibility to political corruption and human rights abuses. However, it became resurgent in a modified form labeled neocommunism, focused on the necessity of government intervention in response to the crisis of the Great Collapse.

cryosphere The portions of the Earth's surface, including glaciers, ice sheets, sea ice, and permafrost on land, that used to be frozen.

environment The archaic concept which, separating humans from the rest of the world, identified the nonhuman component as something which carried particular aesthetic, recreational, or biological value (see *environmental protection*). Sometimes the "natural" environment was distinguished from the "built" environment, contributing to the difficulty that twentieth-century humans had in recognizing and admitting the pervasive and global extent of their impact. Radical thinkers, such as Paul Ehrlich and Dennis and Donella Meadows (a twentieth-century husband and wife team), recognized that humans are part of their environment and dependent upon it, and that its value was more than aesthetic and recreational; that the natural world was essential for human employment, growth, prosperity, and health. These arguments were commonly disparaged, but the idea of environmental protection contained at least partial recognition of this point.

environmental protection The archaic late-twentieth-century concept that singled out the nonhuman environment (see *environment*) for legal protection, typically in response to damaging economic activity (see *external costs*).

eustatic refugees "Eustasy" is a global change in sea level; eustatic refugees are those displaced by rising seas.

external costs In capitalist economic systems (see *capitalism*; *invisible hand*), prices for goods and services were based upon what the market "would bear" (i.e., what consumers were willing and able to pay), without regard to social, biological, or physical costs associated with manufacture, transport, and marketing. These additional costs, not reflected in prices, were referred to as "external" because they were seen as being external to markets and therefore external to the economic system in which those markets operated (see *market failure*). Economists of this era found it difficult to accept that one could not have an economy without the resources provided by this "external" environment.

Fisherian statistics A form of mathematical analysis developed in the early twentieth century and designed to help distinguish between causal and accidental relationships between phenomena. Its originator, R. A. Fisher, was one of the founders of the science of population genetics, and also an advocate of racially-based eugenics programs. Fisher also rejected the evidence that tobacco use caused cancer, and his argument that "correlation is not causation" was later used as a mantra by neoliberals rejecting the scientific evidence of various forms of adverse environmental and health effects from industrial products (see *statistical significance*).

fugitive emissions Leakage from wellheads, pipelines, refineries, etc. Considered "fugitive" because the releases were supposedly unintentional, at least some of them (e.g., methane venting at oil wells) were in fact entirely deliberate. While widely acknowledged by engineers to exist, the impacts of fugitive emissions was minimized by the carbon-combustion complex and its defenders, and thus went largely unaccounted (see *bridge to renewables*; *capitalism*; *external costs*). Some went so far as to insist that because methane was a commercially valuable gas, it was impossible that corporations would allow it to "escape."

Great Depression The period of pervasive market failure, deflation, and unemployment, in the United States and Europe, from 1929 to 1941, separating the First Gilded Age from the Second World War. Brought on by the collapse of unregulated financial markets, it led to widespread questioning of capitalist theory, and for a period of about a half century, the deployment of social-democratic policies in Europe and North America designed to ameliorate the worst social costs of market capitalism. However, by the late twentieth century, the lessons of the Great Depression had been forgotten, many of the protections put in place dismantled, and a veritable frenzy of consumption (driven in large part by fossil fuel combustion) ensued.

greenhouse gases Gases—such as water vapor, carbon dioxide, methane, nitrous oxide, and others—that trap infrared radiation and therefore heat the Earth above the temperature that would prevail in their absence.

human adaptive optimism (1) The belief that there are no limits to human adaptability—that we can either adapt to any circumstances, or change them to suit ourselves. Belief in geoengineering as a climate "solution" was a subset of HAO. (2) The capacity of humans to remain optimistic and adapt to changed circumstances, even in the face of daunting difficulties, and even if the form of "adaptation" required is suffering.

invisible hand A form of magical thinking, popularized in the eighteenth century, that economic markets in a capitalist system were "balanced" by the actions of an unseen, immaterial power, which both ensured that markets functioned efficiently and that they would address human needs. Belief in the invisible hand (sometimes also called the invisible hand of the marketplace) formed a kind of quasi-religious foundation for capitalism (see *capitalism; external costs; market failure; market fundamentalism*).

market failure The social, personal, and environmental costs that market economies imposed on individuals and societies were referred to as "market failures." The concept of market failure was an early recognition of the limits of capitalist theory (see *external costs; invisible hand*).

market fundamentalism A quasi-religious dogma (see *invisible hand*) promoting unregulated markets over all other forms of human socioeconomic organization. During the Penumbra, market fundamentalists tended to deny the existence of market failure, thus playing a key role in the denial of the changes that were already under way and therefore in the catastrophes that ensued.

Period of the Penumbra The shadow of anti-intellectualism that fell over the once-Enlightened techno-scientific nations of the Western world during the second half of the twentieth century, preventing them from acting on the scientific knowledge available at the time and condemning their successors to the inundation and desertification of the late twenty-first and twenty-second centuries.

physical scientists The practitioners in a network of scientific disciplines derived from the eighteenth-century natural philosophy movement. Overwhelmingly male, they emphasized study of the world's physical constituents and processes—the elements and compounds; atomic, magnetic, and gravitational forces; chemical reactions, flows of air and water—to the neglect of biological and social realms and focused on reductionist methodologies that impeded understanding of the crucial interactions between the physical, biological, and social realms.

positivism The intellectual philosophy, promoted in the late nineteenth century by the French sociologist Auguste Comte (but also associated with earlier thinkers such as Francis Bacon and Pierre Simon LaPlace and later thinkers such as Ernst Mach and A. J. Ayer), which stressed that reliable knowledge must be grounded in observation. Statements that could not be tested through observation were considered to be outside the realm of "positive knowledge"— or science—and this included most metaphysical and religious claims. Logical positivists (sometimes also referred to as logical empiricists) stressed the linguistic aspects of this problem and focused on finding theoretically neutral means to articulate observation statements. In the twentieth century, the term was sometimes confusingly associated with the nearly opposite stance: that scientific theories are to be believed come hell or high water. Thus, scientists were sometimes accused of being "positivists" for believing in the truth of their theories, when in fact a true positivist would only believe in the truth of the observations (or observation statements) on which those theories were built.

precautionary principle "First, do no harm," a doctrine supposed to descend from the ancient Greek physician Hippocrates. The basis of all policies designed to protect human life and health.

Sagan effect In 1959, U.S. astronomer Carl Sagan identified the greenhouse effect as the cause of Venus's hotter-than-molten-lead surface temperature; as anthropogenic global warming took hold in the late 2000s, the term *Sagan effect* was used to refer to the runaway greenhouse effect on Earth.

sink Place where wastes accumulated, either deliberately or not. Industrial powers of the twentieth century treated the atmosphere and oceans as sinks, wrongly believing them capable of absorbing all the wastes humans produced, in perpetuity.

statistical significance The archaic concept that an observed phenomenon could only be accepted as true if the odds of it happening by chance were very small, typically taken to be no more than 1 in 20.

synthetic-failure paleoanalysis The discipline of understanding past failure, specifically by understanding the interactions (or synthesis) of social, physical, and biological systems.

termination shock The sudden rise in atmospheric temperature caused by the cessation of efforts to cool the planet.

type I error The conceptual mistake of accepting as true something that is false. Both type I and type II errors are wrong, but in the twentieth century it was believed that a type I error was worse than a type II error.

type II error The conceptual mistake of rejecting as false something that is true. In the twentieth century it was believed that a type I error was worse than a type II error. The rejection of climate change proved the fallacy of that belief.

zero-net-carbon infrastructure The technocomplex of non-fossil energy sources, transportation systems, and carbon sequestration technologies deployed to counterbalance the climate effects of the industrial agriculture necessary to sustain the surviving human population (e.g., like the not-very-successful forests of mechanical atmosphere scrubbers deployed between 2100 and 2170).

Interview with the Authors

1. *How did you originally come to write* The Collapse of Western Civilization (CWC)*? How do you see it in relation to your recent title* Merchants of Doubt?*

Naomi Oreskes (NO): I was invited to write a piece for a special issue of *Daedalus* on social scientific approaches to climate change. The specific invitation was to write on why we (collectively) were failing to respond adequately. At the time, I was pondering why scientists' attempts at communication were proving so conspicuously ineffective, but I was having trouble thinking of how to answer the question without rewriting *Merchants of Doubt* on the one hand, or sounding like a scold on the other. As a

* Interview conducted by Patrick Fitzgerald, Publisher for Life Sciences, Columbia University Press.

historian, I also felt uncomfortable trying to answer such a present-tense question. Then the thought came to me: what will future historians say about us? How will they answer this question? And when Erik didn't object . . .

2. *Naturally, contemplating the collapse of a civilization and the planetary environmental disaster you describe is pretty grim stuff. How did you wrestle with the overall mordant quality of the writing?*

NO: Writing from the perspective of a future historian solved that problem. Viewing things in hindsight gives you emotional distance.

Erik Conway (EC): Yes—and I would add that I'm not sure "mordant" is really the right term. Many people on the activist end of the climate change "debate" seem to believe that climate change will result in human extinction. But that's not what looking back at previous episodes of climate change in human history has to offer. Our social, political, and cultural organizations change in response to climate. Sometimes governance structures survive, sometimes they don't. But people will—although if we continue down our current fossil-fueled path, there will be a lot less of us in a few hundred years. Not that I will live to see it!

To me, CWC is hopeful. There will be a future for humanity, even if one no longer dominated by "Western Culture."

3. *How have fiction and science fiction informed your thinking and your writing in CWC?*

EC: Science fiction has explored issues of climate change and disruption for decades. The first fiction author I'm aware of who looked seriously at the role of ecosystems in human activities was Frank Herbert. His 1965 work *Dune* was a fascinating study of the way a single planet's ecosystem affected an entire galactic empire. The work most personally influential to us was Kim Stanley Robinson's two trilogies—his Mars triology, and his climate change trilogy (the latter is the one we mention in the book). They are both complicated stories, but the basic narrative in the Mars story is conflict over how to change the climate to make it inhabitable for Earthlings, while also trying to deal (badly) with climate disruption on Earth. The climate change trilogy is written in a realistic vein, so some might not consider it science fiction at all, but it is closely based on science, as is our story.

There are numerous advantages to working in a fictional mode. One is that you can address themes in ways that are very difficult for historians, because you are not so strictly limited by sources. Robinson, for example, deals very effectively with various strands of environmental thinking. For a historian to do that, he or she would need access to a very large quantity of material, including business records, and it's very difficult to get access to corporate records. History tends to be skewed toward topics and people who have left extensive, and open, records. Historians also have to stay close to their source material, which sometimes gets in the way of telling a good story. Fiction gives you more latitude, and here we try to use that latitude in interesting and thought-provoking ways, but always with the goal of being true to the facts: true to what science tells us could really happen if we continue with business as usual, and true to what history suggests is plausible. Nothing is invented out of whole cloth.

NO: In my talks, I like to remind folks of Robinson's wonderful line: "The invisible hand never picks up the check." That's market failure in a nutshell. Stan was one of the first people to get the connection between neoliberalism and climate change denial and the first to work that theme into fiction, at least as far as I know.

It's interesting to me that he is very influenced by history of science. I suppose it's the sense I got from his work of a blend created by honoring the factual

constraints of nature—as illuminated by science—
with the creative opportunities offered by fiction—
that helped to inspire this piece. I didn't realize that
when we first started writing, but partway into I did.
I'm also inspired by Margaret Atwood. *Alias Grace*
especially.

4. *One of the features I enjoyed is the satire in your essay.
Can you tell us a little about the "carbon-combustion
complex" and the "Sea Level Rise Denial Bill"?*

NO: *Merchants of Doubt* tells the story of a particular group of people who sowed doubt about climate
change (and several other issues), why they did, and
how. But it left (mostly) unanswered the question
of why selling doubt has been so effective. One part
of the answer had to do with the large network of
people who benefit from the production and use of
fossil fuels. It's not just the fossil fuel companies—
although obviously they lay the crucial foundation—
but it's also the automobile industry, the aerospace
industry, the electrical utilities, the folks who supply
asphalt for roads . . . you get the picture. At the same
time we were writing this piece, Erik and I were also
working on a volume on Cold War science. You can't
talk about the Cold War without talking about the
military-industrial complex. The carbon-combustion
complex was a natural analog.

EC: The carbon-combustion complex, as Naomi describes, was a convenient formulation for dealing with the intersecting political and economic interests surrounding fossil energy production and use, though my own tendency is to refer to them as the fossil industry complex. They'll be extinct once the oil runs out, after all. But Naomi won that argument.

NO: I don't like "fossil industry." It makes it sound like they are selling fossils. Agreed, with any luck one day they *will* be fossils, unless they can transform their business model and become energy companies rather than fossil fuel companies. I truly hope that they can do that. I'm a geologist and I still have friends in the oil industry.

EC: The Sea Level Rise Denial Bill was all too real.[1] The political denial of climate change has gotten so absurd that I've wondered if the denialists are being advised by a comedian. Have The Yes Men gone into consulting?

NO: I agree. This would be funny if it weren't (mostly) true. The supporters of the Sea Level Rise Denial Bill don't call it that, of course, but that's what it is. We figured future historians would call a spade a spade.

5. *Why did you decide to situate the narrator in China in the Second (or Neocommunist) People's Republic?*

NO: The doubt-mongers we wrote about in *Merchants of Doubt* were anti-communists who opposed environmental regulations for fear that government encroachment in the marketplace would become a backdoor to communism. They believed that political freedom was tied to economic freedom, so restrictions on economic freedom threatened political freedom. Their views came out of the Cold War—particularly the writings of Milton Friedman and Friedrich von Hayek—but the essential idea remains a tenet for many people on the right wing of the American political spectrum today. While rarely stated quite this baldly, the reasoning goes like this: Government intervention in the market place is bad. Accepting the reality of climate change requires us to acknowledge the need for government intervention either to regulate the use of fossil fuels or to increase the cost of doing so. So we won't accept the reality of climate change.

Erik and I have pointed out that besides being illogical, this sort of thinking—by delaying action—increases the risk that disruptive climate change will lead to the very sort of heavy-handed interventions that conservatives wish to avoid. Catastrophic natural disasters—particularly those that disrupt food and water supply—are a justification for governments to send in the national guard, commandeer resources, declare martial law, and otherwise suspend democratic processes and interfere with markets. Given this, one can make the argument that authoritarian societies will be more able to handle catastrophic climate change than free ones. So people who care about freedom should want to see early action to prevent catastrophic climate change. Delay increases the risk that authoritarian forms of governance will come out ahead in the end. The nation in which our historian is writing is the Second PRC, because we imagine that after a period of liberalization and democratization, autocratic forces become resurgent in China, justified by the imperative of dealing with the climate crisis.

EC: Chinese civilization has been around a lot longer than Western civilization has and it's survived a great many traumas. While I'm not sure the current government of China is likely to hold up well— the internal tensions are pretty glaring—it's hard to imagine a future in which there's no longer a place called China. And as Naomi explains, authoritarian states may well find it easier to make the changes necessary to survive rapid climate change. With a few exceptions, the so-called liberal democracies are failing to address climate change.

Whatever the surviving states of 2393 are called, though, they'll be writing history from their perspective. So there will be a history of the Penumbra as written from the point of view of Russia, and of the South American confederation, of the Nordo-Scandinavian Union, etc.

6. *In your essay you delineate the period of Western civilization as extending from 1540 to 2093. Can you explain these dates?*

NO: Hmm . . . Erik, can you recall why we started with 1540? Was it something to do with Copernicus?

EC: We chose 1540 for Georg Joachim Rheticus's publication of the *Narratio Prima*, the first published argument for heliocentrism. It was written as the introduction to Copernicus's *De revolutionibus orbium coelestium*, published in 1543. Traditionally, historians have taken this work as marking the beginning of the scientific revolution—although most historians today reject that term—and of the ascendance in Western Europe of natural philosophy, with its commitment to understanding the universe on the basis of physical evidence.

The choice of 2093 is more arbitrary. Neither of us will live to see that year, so it seemed far enough in the future for emotional comfort. On the other hand, there's no question that sea level rise by then will have become significant and obvious to anyone near the coastlines.

NO: 2093 felt close enough that it was scary, but not so close that the year would come and go within our lifetimes and people would say—see you got that wrong! And it seemed like a reasonable estimate for when things will start to go badly if we continue business as usual. Roger Revelle [one of the first scientists to warn about global warming] worried a lot about what would happen by the year 2100. So this places the end of our story just before that inflection point.

7. *Certainly one of the more memorable and pungent themes in your essay is the over-reliance of scientists on the 95 percent confidence interval, before they will make a call on causation or recommend any kind of public policy or action. You are stepping on some pretty sacred toes here. Are you worried about the "slippery slope" argument that abandoning long-cherished standards of statistical significance could lead to crappy science and misguided, even dangerous, policy?*

NO: This is a really big issue. There is a lot I'd like to say about that, but, as you note, we are stepping on hallowed ground. Maybe there is another book there. But no, I am not worried about the slippery slope argument. For one thing, slippery slope arguments are generally illogical. Just because you say some element of a system should be reexamined doesn't mean you are burning down the house. The argument against accepting the scientific evidence of climate change that we tracked in *Merchants of Doubt* was a slippery slope argument: today we control greenhouse gas emissions, tomorrow we give up the Bill of Rights. One of our protagonists said this explicitly, in defense of tobacco: that if we allow the government to control tobacco there's no limit to what the government may try to control next. It's a foolish argument. Actions should be based on their merits or demerits. Anything can be foolish if taken too far. Tobacco kills people, and it is addictive. It makes sense to regulate it, just as it makes sense to regulate heroin. It also makes sense to regulate driving, and air traffic control. But that doesn't mean we should regulate soda. Each question has to be debated on its merits. Regulations that made sense in the past, such as in telecommunications, might not make sense today, and regulations that make sense today may need to be revised or repealed in the future.

The challenge is always to determine what is needed in any given situation. It's the same for science. Scientists have changed their standards in the past, and they will do so again. It's high time we had a serious discussion of where the 95 percent confidence limit came from, and whether it makes sense in the nearly indiscriminate way that it is currently applied.

EC: Plus—we have crappy science and dangerous policy despite the existence of the 95 percent confidence limit convention!

NO: Good point!

EC: That's why journals have retraction processes, for one thing. The 95 percent confidence limit is a choice—just like our choice of 2093 as the end of Western civilization—but it's not entirely arbitrary: it's designed to provide a high hurdle against one specific kind of error. As we explain in the piece, and in *Merchants of Doubt*, it provides no defense against many other kinds of error.

8. *So what about the "precautionary principle"—the idea you describe in CWC, that we must take early action to prevent later disaster? Critics of this notion argue that the precautionary principle is more of a rhetorical and advocacy tool than a responsible way to explore and develop policy options. Do you think a market-based, neoliberal political and economic regime can act with long-term caution?*

EC: The claim is nonsense. In terms of anthropogenic climate change, the precautionary principle is moot. Precautions are taken in advance of damage, not after it has already begun. We have overwhelming evidence that we've already triggered a rapid rate of oceanic and atmospheric warming. We're currently reacting to climate change already in progress, not deploying precautions against warming that might or might not happen in the future.

NO: I love Erik! I was going to say the same thing. The precautionary principle deals with what one should do when there is evidence that something may be a problem, but we're not quite sure, or not sure of the extent of it. We *are* sure that climate change is happening—we already see damage—and we know beyond a reasonable doubt that business as usual will lead to more damage, possibly devastating damage, as our story tells. It's way too late for precaution. Now we are talking about damage control.

EC: Insofar as the larger issue of policy beyond climate change, the argument is still wrong. U.S. law requires pharmaceutical manufacturers to perform toxicity testing on new drugs, which is a form of precaution. But our laws, unlike those of the European Union, do not require testing of industrial chemicals. Why do we deploy the precautionary principle against drugs, but not pesticides or plasticizers? Politics. The chemical industry succeeding in preventing precautionary regulation where the pharmaceutical industry did not.

Here's another example. I live in California. There's a large, active fault known as the San Andreas running within twenty miles of my condo. We have both geological and historical evidence that the San Andreas has caused great earthquakes. Because of that evidence, the state has taken the precaution of imposing seismic design requirements in its current building codes. It also holds emergency response drills—my employer, a private technical college, participates in them too.

Those precautionary requirements cost us all money, by increasing the cost of our structures and in productive time lost from work. Yet we don't know when, or where, the next fault rupture will occur. They're unpredictable.

Is the argument here that we shouldn't take these precautions? I doubt you could pass a referendum in California repealing them. Most of us don't really want our buildings falling down on us in the name of protecting "free markets."

Can we deploy the precautionary principle without strangling the economy? Yes, we do it all the time. Precautions are everywhere. A stop sign is a precaution!

Can a neoliberal regime act with long-term caution? No, because the neoliberal worship of deregulation leads directly to the poisoning of ourselves and the rest of the world.

NO: That might be a bit strong. But Erik's basically right. Neoliberalism in its pure form fails to recognize external costs or to provide a mechanism for preventing future damage. There's no market signal from the future, or from birds and bats and bees (until the damage is so great that we actually see it, for example, in the cost of honey, but even then, most consumers won't know why the cost of honey is rising).

Business and political leaders who have been swayed by the arguments for deregulation need to realize that while the basic idea of invoking competition to good ends is a powerful one, it only works in the full sense when tempered by the need to address market failure and external costs. Neoliberalism is an ideology, and like most ideologies, it hits potholes and speed bumps when put into practice; even Adam Smith recognized that you have to regulate the banks. Climate change is a really, really big pothole. But here's an interesting point to note: von Hayek explicitly invoked pollution as an external cost that can legitimately justify government intervention in the marketplace. I suppose that might be why some people on the right deny that CO_2 is a pollutant . . .

9. *One of the great ironies you describe in CWC is that, ultimately, it is the neoliberal regimes that fail to act in time to avert climate change disasters and it is China, the epitome of the command-and-control political culture, that can make the huge institutional moves to save its population. This scenario is pretty breathtaking speculation! Just how much do you hate the American way of life? What gives you the intellectual chutzpah to make these kinds of projections?*

EC: What is this "American Way of Life" you speak of so blithely? Is it the America of one-room schoolhouses on the prairie? Of small-holders, shopkeepers, and family farms? That's what it meant in 1930. But that "American way of life" is gone, and its departure had nothing to do with us. Its vanishing had a great deal to do with the growth of industrial capitalism and the drive for efficiencies of scale. The funny thing about "efficiencies of scale" is that they tend to concentrate wealth—and therefore power—in the hands of a few, and those few can thwart the will of the majority pretty easily. Theodore Roosevelt called this class of men "malefactors of great wealth" in a 1907 speech in Massachusetts. There are simply more of them now (and they've allowed a handful of women into their club).

The wealthiest businesses the world has ever known are part of the carbon-combustion complex, and they've been enormously successful at preventing most of the liberal democracies from doing anything meaningful about climate change. I don't see any reason to believe they'll suddenly throw in the towel and play nice.

NO: Our story is a call to protect the American way of life before it's too late. Speculative? Of course, but the book is extremely fact-based. All the technical projections are based on current science. Chutzpah? You have to have chutzpah to write any book. Or to stand in a classroom and expect students to listen. Strangely enough, they do, and sometimes they even thank you. Readers, too.

10. *What do you hope that readers take away from your essay?*

EC: Readers tend to take out of a text whatever it was that they brought in. At best we can hope to have helped them think more clearly about the climate of the future.

NO: Hmm . . . you can't predict what your readers will take away. Books are like a message in a bottle. You hope someone will open it, read it, and get the message. Whatever that is.

Notes

1. The Coming of the Penumbral Age

1. See http://www.quaternary.stratigraphy.org.uk/workinggroups/anthropocene/.

2. Ronald Doel, "Constituting the Postwar Earth Sciences: The Military's Influence on the Environmental Sciences in the USA after 1945," *Social Studies of Science* 33 (2003): 535–666; Naomi Oreskes, *Science on a Mission: American Oceanography from the Cold War to Climate Change* (Chicago: University of Chicago Press, forthcoming).

3. Paul Ehrlich *The Population Bomb* (New York: Ballantine Books, 1968). See also "Can a Collapse of Global Civilization be Avoided?," Paul R. Ehrich and Anne H. Ehrlich, *Proc. Royal Society B*, 2013.

4. On the various forms of Chinese population control, see Susan Greenhalgh, *Just One Child: Science and Policy in Deng's China* (Berkeley: University of California Press, 2008).

5. See http://unfccc.int/meetings/copenhagen_dec_2009/meeting/6295.php.

2. The Frenzy of Fossil Fuels

1. Michael Mann, *The Hockey Stick and the Climate Wars: Dispatches from the Front Lines* (New York: Columbia University Press, 2012).

2. See http://www.wired.com/wiredscience/2012/06/bp-scientist-emails/.

3. Seth Cline, "Sea Level Bill Would Allow North Carolina to Stick Its Head in the Sand," *U.S. News & World Report*, June 1, 2012, http://www.usnews.com/news/articles/2012/06/01/sea-level-bill-would-allow-north-carolina-to-stick-its-head-in-the-sand. Stephen Colbert made a satire of the law (see Stephen Colbert, "The Word—Sink or Swim," *The Colbert Report*, June 4, 2012, http://www.colbertnation.com/the-colbert-report-videos/414796/june-04-2012/the-word-sink-or-swim).

4. Government Spending Accountability Act of 2012, 112th Cong., 2012, H.R. 4631, http://oversight.house.gov/wp-content/uploads/2012/06/WALSIL_032_xml.pdf.

5. Kim Stanley Robinson, *Forty Signs of Rain, Fifty Degrees Below, and Sixty Days and Counting* (New York: Spectra Publishers, 2005–2007).

6. Naomi Oreskes, "Seeing Climate Change," in *Dario Robleto: Survival Does Not Lie in the Heavens*, ed. Gilbert Vicario (Des Moines, Iowa: Des Moines Art Center, 2011).

7. See Clive Hamilton, *Requiem for a Species: Why We Resist the Truth about Climate Change* (Sydney: Allen and Unwin, 2010), http://www.clivehamilton.net.au/cms/; and Paul Gilding, *The Great Disruption: Why the Climate Crisis Will Bring On the End of Shopping and the Birth of a New World* (New York: Bloomsbury Press, 2010).

8. For an electronic archive of predictions and data as of 2012, see http://www.columbia.edu/~mhs119/Temperature/T_moreFigs/.

An interesting paper from the University of California, San Diego, addresses the issue of under-prediction; see Keynyn Brysse et al., "Climate Change Prediction: Erring on the Side of Least Drama?" *Global Environmental Change* 23 (2013): 327–337.

9. David F. Noble, *A World Without Women: The Christian Clerical Culture of Western Science* (New York: Knopf, 1992); and Lorraine Daston and Peter L. Galison, *Objectivity* (Cambridge, Mass.: Zone Books, 2007).

10. Naomi Oreskes and Erik M. Conway, *Merchants of Doubt: How a Handful of Scientists Obscured the Truth on Issues from Tobacco to Climate Change* (New York: Bloomsbury, 2010), chap. 5, esp. 157 n. 91–92. See also Aaron M. McCright and Riley E. Dunlap, "Challenging Global Warming as a Social Problem: An Analysis of the Conservative Movement's Counter-claims," *Social Problems* 47 (2000): 499–522; Aaron M. McCright and Riley E. Dunlap, "Cool Dudes: The Denial of Climate Change among Conservative White Males in the United States," *Global Environmental Change* 21 (2011) 1163–1172.

11. Justin Gillis, "In Poll, Many Link Weather Extremes to Climate Change," *The New York Times*, April 17, 2012, http://www.nytimes.com/2012/04/18/science/earth/americans-link-global-warming-to-extreme-weather-poll-says.html.

12. Tom A. Boden, Gregg Marland, and Robert J. Andres, "Global, Regional, and National Fossil-Fuel CO_2 Emissions," Carbon Dioxide Information Analysis Center (Oak Ridge, Tenn.: Oak Ridge National Laboratory, 2011), http://cdiac.ornl.gov/trends/emis/overview_2008.html.

13. Sarah Collins and Tom Kenworthy, "Energy Industry Fights Chemical Disclosure," Center for American Progress, April 6, 2010, http://www.americanprogress.org/issues/2010/04/fracking.html; Jad Mouawad, "Estimate Places Natural Gas Reserves 35% Higher," *The New York Times*, June 17, 2009,

http://www.nytimes.com/2009/06/18/business/energy-
environment/18gas.html?_r=1.

14. See http://www.eia.gov/naturalgas.

15. Emil D. Attanasi and Richard F. Meyer, "Natural Bitumen
and Extra-Heavy Oil," in *Survey of Energy Resources*, 22nd ed.
(London: World Energy Council, 2010), 123–140.

16. David W. Schindler and John P. Smol, "After Rio, Canada
Lost Its Way," *Ottawa Citizen*, June 20, 2012, http://www.
ottawacitizen.com/opinion/op-ed/Opinion/6814332/story.html.

17. See http://security.blogs.cnn.com/2012/06/08/militarys-
plan-for-a-green-future-has-congress-seeing-red/.

18. "Georgia Power Opposes Senate Solar Power Bill," *The
Augusta Chronicle*, February 18, 2012, http://chronicle.augusta.
com/news/metro/2012–02–18/georgia-power-opposes-senate-
solar-power-bill.

19. Arctic Sea Ice Extent, IARC-JAXA Information System
(IJIS), accessed October 10, 2013: http://www.ijis.iarc.uaf.edu/
en/home/seaice_extent.htm; Arctic Sea Ice News and Analysis,
National Snow & Ice Data Center, accessed October 10, 2013:
http://nsidc.org/arcticseaicenews/; Christine Dell'Amore, "Ten
Thousand Walruses Gather on Island As Sea Ice Shrinks," *National
Geographic*, October 2, 2013; William M. Connolley, "Sea ice
extent in million square kilometers," accessed October 10, 2013:
http://en.wikipedia.org/wiki/File:Seaice-1870-part-2009.png.

20. Gerald A. Meehl and Thomas F. Stocker, "Global Climate
Projections," in *Fourth Assessment Report of the Intergovernmental
Panel on Climate Change*, "Climate Change 2007—The Physical
Science Basis." February 2, 2007.

21. Clifford Krauss, "Exxon and Russia's Oil Company in Deal
for Joint Projects," *The New York Times*, April 16, 2012.

22. For statistics on continued coal and oil use in the mid-
twentieth century, see U.S. Energy Information Administration, *International Energy Outlook 2011* (Washington, D.C.:

U.S. Department of Energy, 2011), 139, Figures 110–111, http://205.254.135.7/forecasts/ieo/.

23. On twentieth- and twenty-first-century subsidies to fossil fuel production, see http://www.oecd.org/document/57/0,3746,en_2649_37465_45233017_1_1_1_37465,00.html; and John Vidal, "World Bank: Ditch Fossil Fuel Subsidies to Address Climate Change," *The Guardian*, September 21, 2011, http://www.guardian.co.uk/environment/2011/sep/21/world-bank-fossil-fuel-subsidies.

24. "Canada, Out of Kyoto, Must Still Cut Emissions: U.N.," Reuters, December 13, 2011, http://www.reuters.com/article/2011/12/13/us-climate-canada-idUSTRE7BC2BW20111213; Adam Vaughan, "What Does Canada's Withdrawal from Kyoto Protocol Mean?" *The Guardian*, December 13, 2011, http://www.guardian.co.uk/environment/2011/dec/13/canada-withdrawal-kyoto-protocol; James Astill and Paul Brown, "Carbon Dioxide Levels Will Double by 2050, Experts Forecast," *The Guardian*, April 5, 2001, http://www.guardian.co.uk/environment/2001/apr/06/usnews.globalwarming.

25. Acknowledgments to http://www.epsrc.ac.uk/newsevents/news/2012/Pages/spiceprojectupdate.aspx.

26. Paul Crutzen, "Albedo Enhancement by Stratospheric Sulfur Injections: A Contribution to Resolve a Policy Dilemma?," *Climatic Change* 77 (2006): 211–219, http://www.springerlink.com/content/t1vn75m458373h63/fulltext.pdf. See also Daniel Bodansky, "May We Engineer the Climate?," *Climatic Change* 33 (1996): 309–321. Also see http://www.handsoffmotherearth.org/hose-experiment/spice-opposition-letter/.

27. Andrew Ross and H. Damon Matthews, "Climate Engineering and the Risk of Rapid Climate Change," *Environmental Research Letters* 4(4) (2009), http://iopscience.iop.org/1748-9326/4/4/045103/.

28. Ian Allison, et al., *The Copenhagen Diagnosis: Updating the World on the Latest Climate Science* (Sydney: University of

New South Wales Climate Change Research Centre, 2009), esp. 21; Jonathan Adams, "Estimates of Total Carbon Storage in Various Important Reservoirs," Oak Ridge National Laboratory, http://www.esd.ornl.gov/projects/qen/carbon2.html.

29. See http://www.sciencedaily.com/releases/2012/03/120312003232.htm; http://www.pnas.org/content/105/38/14245.short.

30. Philip Ziegler, *The Black Death* (London: The Folio Society, 1997).

31. A. Hallam and P. B. Wignall, *Mass Extinctions and their Aftermath* (NY: Oxford University Press, 1997), gives the Big 5 as occurring at the end of the Devonian, Ordovician, Permian, Triassic, and Cretaceous periods, on the classical Western geological timescale.

3. Market Failure

1. Amory Lovins, *Reinventing Fire: Bold Business Solutions for the New Energy Era* (White River Junction, Vt.: Chelsea Green, 2011).

2. Naomi Oreskes and Erik M. Conway, *Merchants of Doubt: How a Handful of Scientists Obscured the Truth on Issues from Tobacco to Climate Change* (New York: Bloomsbury, 2010).

3. For an example of this characteristic, see Justin Gillis, "Rising Sea Levels Seen as Threat to Coastal U.S.," *The New York Times*, March 13, 2012, http://www.nytimes.com/2012/03/14/science/earth/study-rising-sea-levels-a-risk-to-coastal-states.html. Note how Gillis frames the evidence, first stating that "the handful of climate researchers who question the scientific consensus about global warming do not deny that the ocean is rising. But they often assert that the rise is a result of natural climate variability." He then quotes Myron Ebell, who was not a climate researcher, but

an *economist* and paid employee of the Competitive Enterprise Institute, a think tank that was heavily funded by the carbon-combustion complex and committed to market fundamentalism. See http://cei.org/.

4. Richard Somerville, *The Forgiving Air: Understanding Environmental Change* (Washington, DC: American Meteorological Society, 2008); Stephen H. Schneider, *Science as a Contact Sport: Inside the Battle to Save the Earth's Climate* (Washington DC: National Geographic Press, 2009); Gavin Schmidt and Joshua Wolfe, *Climate Change: Picturing the Science* (New York: W. W. Norton, 2009); James Hansen, *Storms of My Grandchildren* (New York: Bloomsbury Press, 2010); Burton Richter, *Beyond Smoke and Mirrors: Climate Change and Energy in the 21st Century* (Cambridge: Cambridge University Press, 2010); Michael Mann, *The Hockey Stick and the Climate Wars: Dispatches from the Front Lines* (New York: Columbia University Press, 2012). For an analysis of scientists' difficulties in dealing efficaciously with public communication media, see Maxwell T. Boykoff, *Who Speaks for the Climate? Making Sense of Media Reporting on Climate Change* (Cambridge: Cambridge University Press, 2011).

5. Richard White, *Railroaded: The Transcontinentals and the Making of Modern America* (New York: W. W. Norton, 2011).

6. Scholars note that even nineteenth-century markets were not free. See Ha-Joon Chang, *Bad Samaritans: The Myth of Free Trade and the Secret History of Capitalism* (New York: Bloomsbury, 2008); and Ha-Joon Chang, *23 Things They Don't Tell You about Capitalism* (New York: Bloomsbury, 2012).

7. Dennis Tao Yang, "China's Agricultural Crisis and Famine of 1959–1961: A Survey and Comparison to Soviet Famines," *Comparative Economic Studies* 50 (2008): 1–29.

8. George H. W. Bush, "Remarks on Presenting the Presidential Medal of Freedom Awards," November 18, 1991.

9. Naomi Oreskes, "Science, Technology, and Free Enterprise," *Centaurus* 52 (2011): 297–310; and John Krige, *American Hegemony and the Postwar Reconstruction of Science in Europe* (Cambridge, Mass.: MIT Press, 2006).

10. See, for example, David Joravsky, *The Lysenko Affair* (Chicago: University of Chicago Press, 1986); and Nils Roll-Hansen, *The Lysenko Effect* (Amherst, N.Y.: Humanity Books, 2004).

11. On twentieth- and twenty-first-century subsidies for fossil fuel production, see http://www.oecd.org/document/57/0, 3746,en_2649_37465_45233017_1_1_1_37465,00.html; and John Vidal, "World Bank: Ditch Fossil Fuel Subsidies to Address Climate Change," *The Guardian*, September 21, 2011, http://www.guardian.co.uk/environment/2011/sep/21/world-bank-fossil-fuel-subsidies.

12. Friedrich August von Hayek, *The Road to Serfdom, Text and Documents: The Definitive Edition*, ed. Bruce Caldwell (Chicago: University of Chicago Press, 2007), 87.

Epilogue

1. For estimates of populations at or near sea level at the turn of the twenty-first century, see Don Hinrichsen, "The Coastal Population Explosion," in *The Next 25 Years: Global Issues*, prepared for the National Oceanic and Atmospheric Administration Coastal Trends Workshop, 1999, http://oceanservice.noaa.gov/websites/retiredsites/natdia_pdf/3hinrichsen.pdf; and http://oceanservice.noaa.gov/websites/retiredsites/supp_natl_dialogueretired.html.

Interview with the Authors

1. See http://www.earthmagazine.org/article/denying-sea-level-rise-how-100-centimeters-divided-state-north-carolina.

About the Maps Inspiration for the 20-meter sea level rise in our scenario came from the Ohio State glaciologist John H. Mercer, who perceived the possibility of a rapid disintegration of the West Antarctic Ice Sheet in 1968 and published a detailed examination in 1978.[*] Recent satellite observations suggest Mercer was more right than he knew, and rapid ice loss is occurring in both West Antarctica and Greenland.[†] The maps drew on Shuttle Radar Topography Mission data from the Jet Propulsion Laboratory at the California Institute of Technology.[‡]

[*]John H. Mercer, "West Antarctic Ice Sheet and CO2 Greenhouse Effect: A Threat of Disaster," *Nature* 271:5643 (1978): 321–25, DOI:10.1038/271321a0.

[†] Andrew Shepherd et.al., "A Reconciled Estimate of Ice-Sheet Mass Balance," *Science* 338 (2012): 1183, DOI:10.1126/science.1228102.

[‡] T. G. Farr et al., "The Shuttle Radar Topography Mission," *Rev. Geophysics* 45 (2007), RG2004, DOI:10.1029/2005RG000183.

About the Authors

Naomi Oreskes is Professor of the History of Science at Harvard University, and affiliated Professor of Earth and Planetary Sciences. Her books include *The Rejection of Continental Drift: Theory and Method in American Earth Science* (1999), *Merchants of Doubt: How a Handful of Scientists Obscured the Truth on Issues from Tobacco to Global Warming* (with Erik M. Conway, 2010), and *Science on a Mission: American Oceanography from the Cold War to Climate Change* (forthcoming). Her latest project is *Assessing Assessments: A Historical and Philosophical Study of Scientific Assessments for Environmental Policy in the Late 20th Century* (with Michael Oppenheimer, Dale Jamieson, Jessica O'Reilly, Matthew Shindell, and Keynyn Brysse).

Erik M. Conway is a historian of science and technology based in Pasadena, California. His publications include *Blind Landings: Low-Visibility Operations in American Aviation, 1918–1958* (2006), *Atmospheric Science at NASA: A History* (2008), and *Merchants of Doubt: How a Handful of Scientists Obscured the Truth on Issues from Tobacco to Global Warming* (with Naomi Oreskes, 2010). His next book will be the history of NASA's Jet Propulsion Laboratory and the Exploration of Mars (forthcoming).